PIRSCH FIRE APPARATUS
1890 THROUGH 1991
PHOTO ARCHIVE

Roger E. Bjorge &
Ronald P. Doerring

Iconografix
Photo Archive Series

Iconografix
PO Box 446
Hudson, Wisconsin 54016 USA

Library of Congress Card Number: 2002104774

ISBN 1-58388-082-8

02 03 04 05 06 07 08 5 4 3 2 1

Printed in China

Cover and book design by Shawn Glidden

Copyediting by Suzie Helberg

COVER PHOTO: The days of retirement were already "in the air" when this 1968 Pirsch Model 41D, Serial No. 2885, had its picture taken on a cold day in March 2000 in Jefferson, Wisconsin. Chief of Department, John Powell, said that a brand new Pierce aerial ladder would soon be taking its place. The bell that was once mounted on the front corner originally came from the city's 1924 Sterling/Pirsch ladder truck. It was already mounted in a prominent place on the new ladder truck.

BOOK PROPOSALS

Iconografix is a publishing company specializing in books for transportation enthusiasts. We publish in a number of different areas, including Automobiles, Auto Racing, Buses, Construction Equipment, Emergency Equipment, Farming Equipment, Railroads & Trucks. The Iconografix imprint is constantly growing and expanding into new subject areas.

Authors, editors, and knowledgeable enthusiasts in the field of transportation history are invited to contact the Editorial Department at Iconografix, Inc., PO Box 446, Hudson, WI 54016.

DEDICATION

This book is dedicated to all the firefighters who serve to protect, and to the workers at Peter Pirsch who labored to build some of the world's finest firefighting apparatus and equipment.

ACKNOWLEDGMENTS

Most of the photographs are copies of originals from the Peter Pirsch & Sons, Incorporated Archives, and are from the collections of the authors. Historic information is also gleaned from Pirsch archives, interviews with former workers at the plant, the Kenosha Historical Society, and Kenosha Simmons Library records. The authors wish to express sincere gratitude to Kerry Poltrock, present owner of the Pirsch name and archives, for his invaluable assistance in completing this project. Other photographs were taken by well-known fire apparatus photographers Richard M. Adelman, Paul Barrett, Bill Friedrich, Matthew Lee, Walter P. McCall, Larry Phillips, George Reichhardt, Steve Skaar, John Sytsma, and Robert James Vigliaturo, whose assistance is much appreciated. Many others who helped greatly were Jack Connors, Gene Brotzman, Cait Dallas, Tommy Fox, Robert Fuhrman, Carl Groth, John Gumbinger, William Kaeppeler, Florian Kreft, Donald Leiting, Faith MacDade, Dan Martin, Charles Madderom, George Mich, Walter Schmidt, Mark Stampfl, The Peter Pirsch Company, The Kenosha Historical Society, and The Kenosha Gilbert M. Simmons Library. We must thank Susan Bjorge for her long hours at the computer assisting us in the completion of this book. We hope we have not left anyone out—if we did, please accept our apologies. This book could not have been done without the input of many others through the years.

Roger E. Bjorge & Ronald P. Doerring

BIBLIOGRAPHY

Walter P. McCall — *American Fire Engines Since 1900*
Matthew Lee — *A Pictorial History of the Fire Engine, Volumes 1 & 2*

This artist's birds-eye view of the Pirsch plant, looking in a northwesterly direction, shows both the original building (left) built in 1916 and the addition (right) added on in 1920. The new addition was built to match the existing one. Both measured approximately 100x200 feet. The plant built at 35th Street in Kenosha paralleled the Chicago & Northwestern Railroad tracks you see in the foreground. The loading dock was used to ship fire apparatus by rail to points all over the U.S. as well as overseas.

INTRODUCTION

PETER PIRSCH & SONS, INC.

Nicholas Pirsch arrived from Luxembourg, Germany in 1850. He settled in Kenosha, Wisconsin in 1856 and started building wagons for the Mitchell Wagon Works in Racine, Wisconsin. In 1861, he opened the Nicholas Pirsch Carriage and Wagon Works. A son, Peter, was born in Kenosha on March 2, 1866, and at age 14, Peter left school to work for his father as a wagon maker and blacksmith. After the disastrous Simmons Company fire in the spring of 1892, which taxed the volunteer fire companies to their limits, Peter resolved that his home town firefighters would be better equipped in the future.

The Pirsch family started making hand-drawn and horse-drawn fire apparatus. In 1895, Peter Pirsch developed the first horse-drawn Hose Cart for Kenosha, followed by a City Service Hook & Ladder later that year. Both carts were built by the Nicholas Pirsch Wagon Works and were each pulled by two horses. Peter applied for a patent on March 19, 1898, for a new Wooden Trussed Ladder. Patent No. 621,427 was granted March 21, 1899, for a Peter Pirsch Extension Ladder. In 1899, Peter Pirsch joined the Kenosha Volunteer Fire Department at age 33, as a member of the James S. Barr Hook & Ladder Company No. 1.

Peter Pirsch left his father's business in 1900 to start his own business, the Peter Pirsch Company, to build his patented Trussed Fire Ladder and Fire Apparatus. He started his business with his patent and thirteen dollars capital. Peter Pirsch invented a Wire Cutting Tool in 1901, and the Pirsch patented Hose Shut Off and Door Opener Tool in 1908. Also, in 1908, Peter Pirsch joined with the Jeffery Automobile Company of Kenosha (later to become Nash Motors) in building the first Motor-driven Fire Truck in the state of Wisconsin, and one of four in existence in the country. In 1910, Peter's son Edward joined his father in the business.

The Pirsch Company built many different types of pumpers, chemical, and ladder trucks on various makers' chassis. In 1913, Peter Pirsch built Kenosha its first Pirsch Motorized City Service Ladder Truck, and in 1916 Creston, Iowa received the first Pirsch equipped with a Triple-Combination pumper. During 1916, Peter Pirsch also invented the Peter Pirsch patented Folding Tongue Rest for Municipal Chemical Engines.

After moving to different locations a few times, a new factory was built in 1916. It was located on 35th Street by the Chicago & Northwestern tracks. A second building was added in 1920. Peter Pirsch, his son, William R., and Celia Pirsch even entered into a co-partnership in 1919 and the company name was changed to Peter Pirsch and Sons Company.

During World War I and World War II, the Pirsch Company built different types and sizes of fire engines for the U.S. Army and U.S. Navy, and supplied many special Pirsch 65-foot ladders to the U.S. Navy to repair damaged blimps.

The first Pirsch-built chassis was developed in 1926, and first used with the new Pirsch 500-gpm triple-combination pumper, followed by the 600-gpm pumper that same year. The 750-gpm pump followed in 1927, the 1,000-gpm pump in 1928, and soon the 1,250-gpm and 1,500-gpm pumpers. The first Pirsch 1,000-gpm triple-combination pumper was delivered to the city of Fond du Lac, Wisconsin in 1925. The Peter Pirsch Smoke Ejector Machine saved 17 men in a tunnel fire in Chicago, Illinois in 1931 while the truck was still under construction. The world's first all-powered Hydro-Mechanical Aerial Ladder Truck was built in 1931 and was delivered to Spokane, Washington. Another ladder "first" was the Pirsch Aluminum-Alloy Aerial Ladder; the first American-built metal aerial ladder longer than 85-feet in length, which was delivered to the city of Melrose, Massachusetts in 1936. The wooden aerial and ground ladders were quickly replaced by the new aluminum-alloy, and the last five tractor-drawn wooden aerial ladder trucks went to Boston, Massachusetts.

In 1946, the Peter Pirsch and Sons Company incorporated and the name changed to Peter Pirsch and Sons, Inc. Peter Pirsch died suddenly in his original workshop on July 14, 1954. The first Pirsch Snorkel was built for Cedar Rapids, Iowa in 1963 using a Pitman Snorkel boom.

Through the years, Pirsch fire trucks were highly regarded by firefighters around the world. This plant, owned and operated by people from Kenosha, was in business in town for 129 years. During the mid-1970s, the Pirsch Company ceased the manufacturing of ground ladders and concentrated on the aerial ladder business. Kenosha, Wisconsin received its last Pirsch pumper, a 1984 Custom 1,250-gpm diesel-powered pumper. Always a Pirsch town, this would be the last Pirsch truck they would be able to buy.

The new Pirsch Skytop 110 Aqua with a 110-foot ladder and a remote-controlled Telescoping Waterway was built in April 1984, and went to Wilbraham, Massachusetts. During 1984, citing labor problems the Pirsch Company purchased 11.4 acres of land in a new Hampton, Virginia Industrial Park with the intention of building a new factory.

However, President Andrew Sale ordered the doors shut and the plant closed down on September 10, 1986. On September 30, 1986, the Pirsch Company filed for Chapter 11 Re-organizational Bankruptcy. At this time there were four or five trucks nearly completed, and sixteen in various stages of completion. Other companies were also building Pirsch fire apparatus for the company at this time. In December 1986 an employee group and some investors attempted to purchase the plant but missed the deadline imposed by the court. The plant was sold to investors John Blondek, and Edward and John Haas of Racine, Wisconsin.

The plant re-opened for business on February 11, 1987, as The Pirsch Company. During February 1987 Edward Haas bought out his other partners and was now sole owner of The Pirsch Company, as well as President and Chief Executive. The first order of business was completing a Redi-Tower pumper for Elizabeth, New Jersey.

The Blondek/Haas team tried to get the business up and running like it used to. Once again in June 1989 Pirsch closed its doors; this time for inventory and in preparation for selling to some Illinois investors. Creditors tried to force The Pirsch Company into involuntary bankruptcy, and in October 1989, the Illinois investors decided they were no longer interested in purchasing the company. It was then that M&I Bank of Milwaukee, owners of The Pirsch Company mortgage, sought foreclosure on the property. Local businessmen eventually purchased the plant and property in April 1991.

The main plant was rented to the Sutphen Corporation for the purpose of building fire engines, and was known as Sutphen West. Shortly thereafter the plant closed. Sutphen had purchased the last Kenosha Pirsch tractor-drawn aerial truck, without the ladders.

The Pirsch office building, ladder shop, and parking lot was then purchased by North Central Fire Apparatus of Kenosha, Wisconsin. Their first job was to complete the last Pirsch fire truck started by the Haas/Blondek Pirsch Company. This pumper was delivered to Osceola, Arkansas in October 1991.

North Central Fire Apparatus is now North Central Emergency Equipment, a Midwest dealer of National Emergency Vehicles located in Orlando, Florida. They also re-manufacture Pirsch aerial ladders, specializing in aerial hydraulics for Pirsch and other makes, and still operate out of these buildings. Kerry Poltrock, owner of NCEV also purchased the name Pirsch and all rights to and contents of The Pirsch Company. The name Pirsch still lives on in Kenosha, Wisconsin.

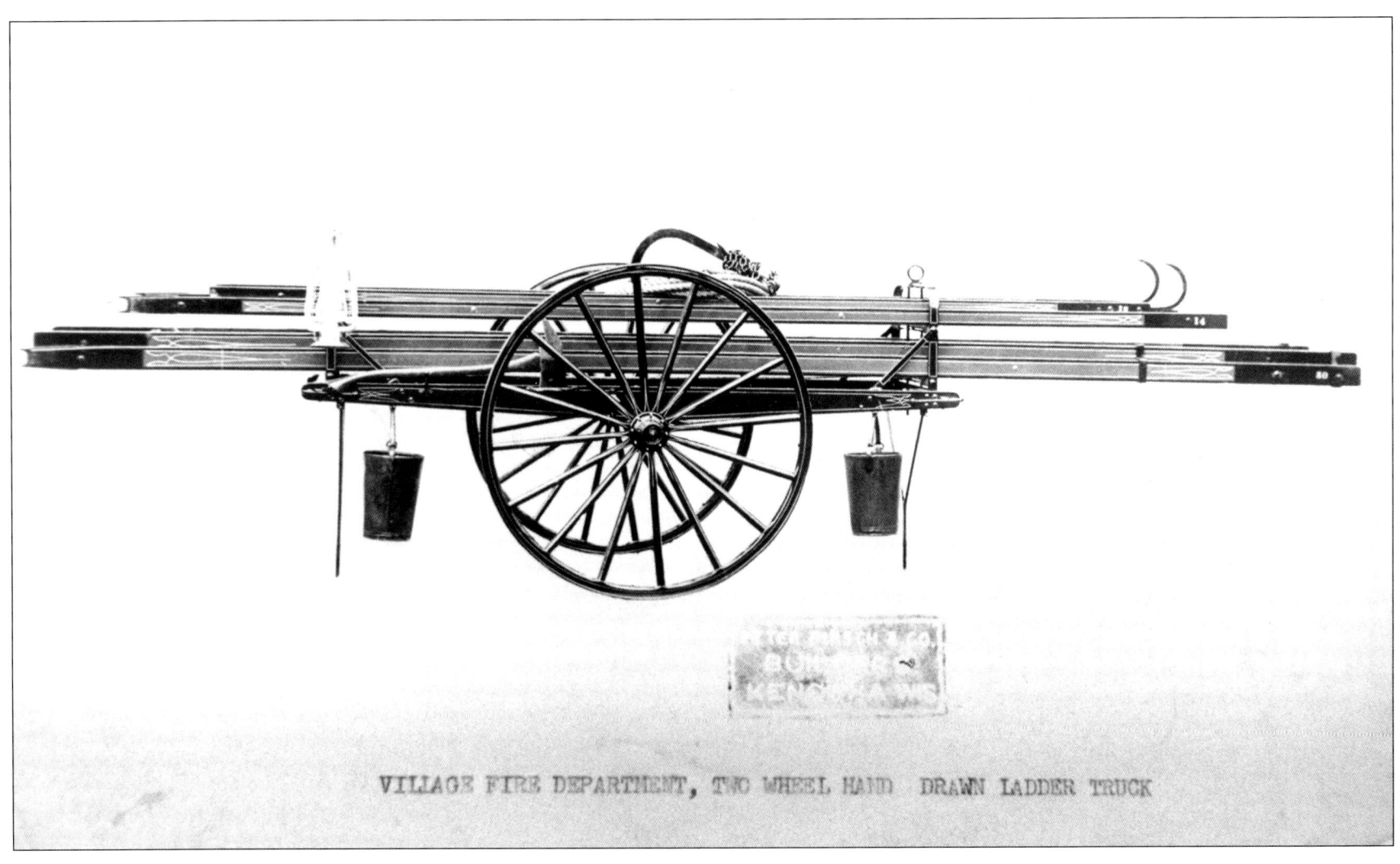

VILLAGE FIRE DEPARTMENT, TWO WHEEL HAND DRAWN LADDER TRUCK

Pirsch assigned names and numbers to describe the various types of fire equipment they were building. This "Village Fire Department, Two Wheel Hand-Drawn Ladder Truck" was provided for villages and small towns that did not require the larger 4-wheel ladder wagons. A 12-foot roof, 14-foot straight or wall, and a 30-foot ladder are carried on this hand-drawn cart, as well as a lantern, an axe and two buckets.

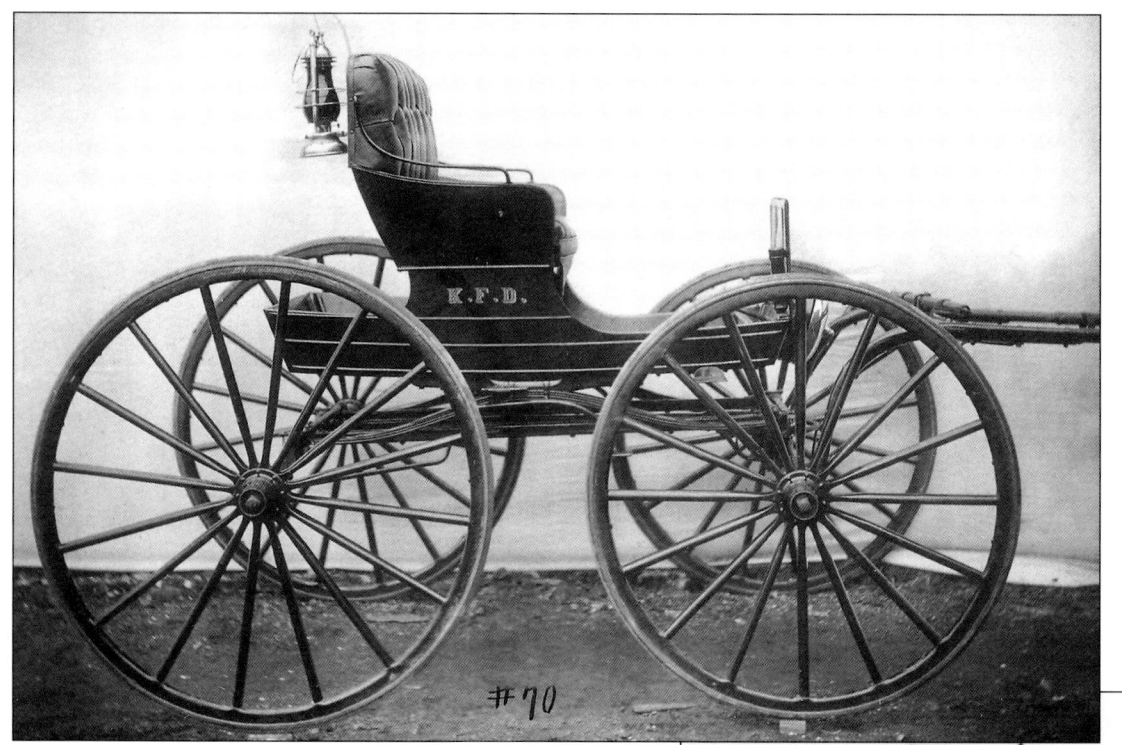

In the 1890s Nicholas Pirsch & Sons built wagons and carriages of every description—both hand and horse drawn. Fire Chief's buggies, such as the one built for Kenosha, were popular. Many of them were built, with a large order having been placed by the Chicago Fire Department.

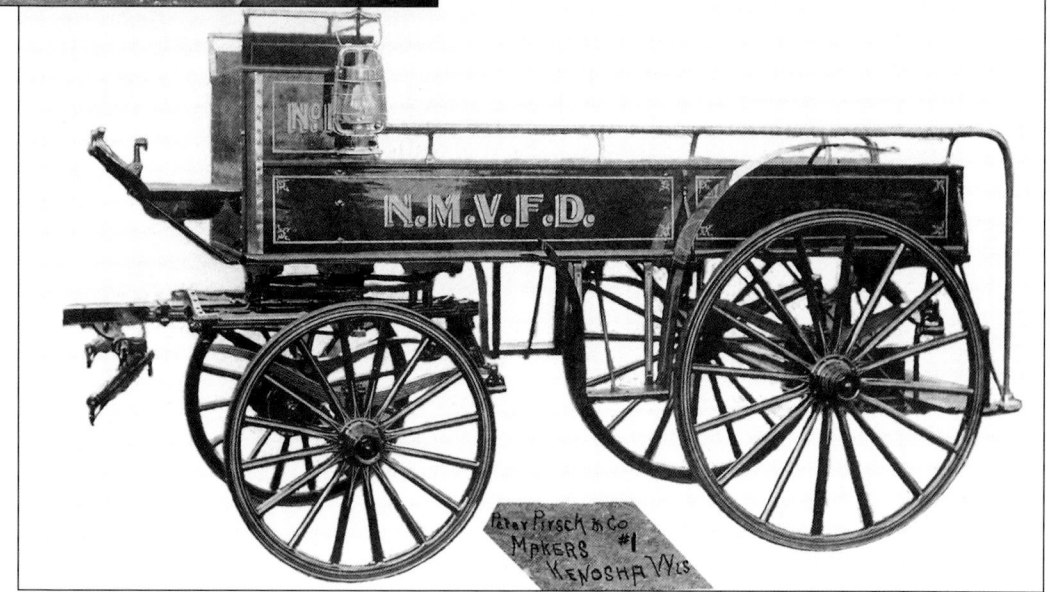

In the late 1890s and early 1900s, a large number of orders came in for horse-drawn wagons such as this. Wagons of this type were used to carry hoses, ladders, tools and equipment, as well as coal for supplying the steamers at fires.

By 1895, Peter Pirsch had established a solid reputation for building quality ladders incorporating the famous trussed design, as well as wagons and carriages of most types. His company's expansion into the larger version of the ladder wagons began with this 55-foot horse-drawn ladder wagon built for the Kenosha Fire Department. This wagon belonged to the James S. Barr Ladder Company No. 1, of which Peter Pirsch was a member. In the construction of Pirsch ladders of both the trussed and solid-side type only the most carefully selected materials were used. The sides and rails were made of Douglas Fir. Rungs were turned from second growth hickory or ash and were allowed to dry naturally before using. Most of the forged iron used was hand forged, rope was manila, and the ladder locks were of Pirsch's own design.

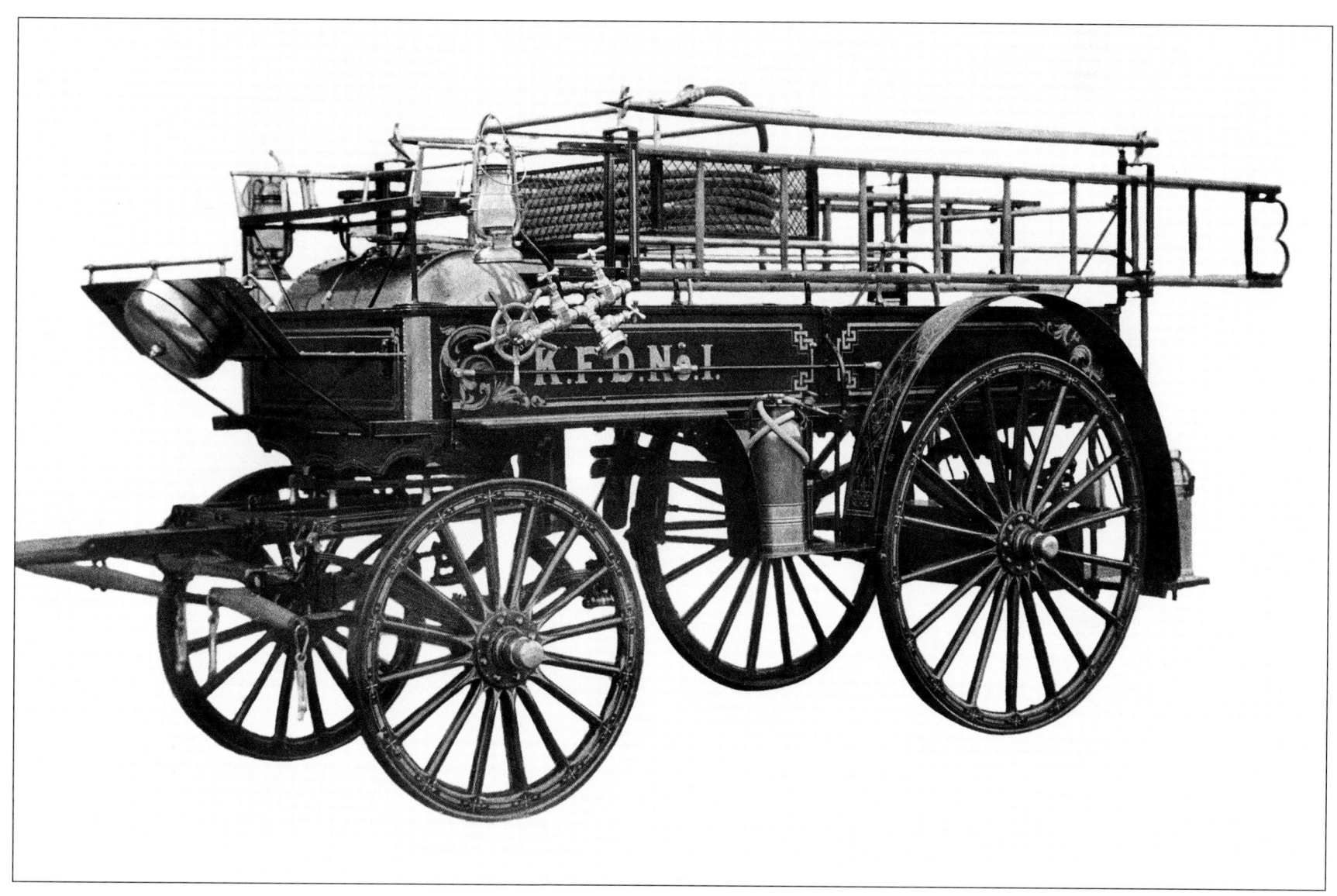

This triple-combination chemical, hose and ladder wagon was built for Kenosha in 1904 by Nicholas Pirsch & Sons, where Peter as a young lad while working under his father's expertise, continued to build a variety of wagons—including fire equipment—for use by people of other trades. This horse-drawn combination shows how advanced the company had become in the field of building quality fire equipment.

PIRSCH VILLAGE TYPE HAND DRAWN CHEMICAL FIRE ENGINE

The hand-drawn chemical "engine" shown here (circa 1910), was manufactured in a variety of sizes and capacities to suit the customer's needs. Tank size varied from 40 to 60 gallons. The chemical hose provided was of 3/4-inch and 100-feet long. Prices ranged from $465 for a 40-gallon cart to $515 for the 60-gallon cart. Orders for hand-drawn hose carts and chemical engines were received from customers all over the United States and Canada, as well as from Central and South America.

In 1908, Peter Pirsch joined with the Jeffery Automobile Company (later known as Nash Motors) in building the first motor-driven fire truck in the state of Wisconsin, and one of four in existence in the United States. Jeffery built the chassis. Pirsch made the body and ladders, and added the chemical equipment. The city of Kenosha received this truck in exchange for vacating a street so that Jeffery could enlarge its plant.

KENOSHA'S FIRST MOTORIZED LADDER TRUCK

DELIVERED AUGUST 27, 1913

Kenosha's ladder companies were either horse-pulled or hand drawn. In 1913, the Kenosha Fire Department had Pirsch assemble some "hook and ladder equipments" on an Atterbury chassis. The first motorized truck had a set of ladder arches that had the famous Pirsch trussed ladders stacked above and below each other as well as ladders that hung on the sides. A chemical tank with a basket for the storage of chemical hose was also included. The Atterbury Motor Car Company of Buffalo, New York made Atterbury trucks. Records reveal that Pirsch assembled at least one more "hook and ladder equipments" on the Atterbury chassis for Aberdeen, Washington.

Order No. 1556: The city of Duluth, Minnesota was a major Great Lakes shipping port for resources such as iron ore, wood and coal. The city's growth required the purchase of this 1916 White city service hook and ladder truck shown here in front of the new Fire Hall No. 8 built the same year. The longest extension ladder carried on this rig was a 55-foot ladder with poles. It also had a 40-gallon "Pirsch" chemical tank located behind the gasoline tank. Other equipment included wire cutters and a life net. *Photo Credit: Robert J. Vigliaturo*

Order No. 1862: It is believed that in 1917 Nokomis, Illinois received this Ford Model T. Shown here is a combination chemical and hose unit. The "Champion" chemical tank could supply 35 gallons of fire-killing chemical through 200 feet of 3/4-inch 4-ply chemical hose. Because of the Model T, fundamental changes were implemented in the way fire trucks were produced. Simple in design, they were readily adapted into almost any type of firefighting unit. Because of this, the ubiquitous "T" could be found in many firehouses across the nation.

This 1918 triple-combination Winther must have "felt right at home" because it was built in the same town that it served—Kenosha, Wisconsin. The Winther proved to be a sturdy and reliable chassis for the rough service that was demanded of it. Because of this, the fire department ordered a ladder truck and another pumper, both on the Winther chassis. This pumper had a 40-gallon chemical tank and a 500-gpm Rumsey pump. Note the combination hose shut-off and door opener mounted on the running board. This was Pirsch's own invention.

Order No. 3812: This rugged, yet handsome looking, four-wheel drive Oshkosh triple-combination unit was one of three identical pumpers built in 1921 for Oshkosh, Wisconsin. The rig is a real showpiece. A large number of fittings, screws, bolts, etc., was specified in the shop order to be nickel-plated. The 40-gallon chemical tank has "hammered" ends for additional fancywork. All ground ladders were of Pirsch's famous trussed design.

The village of Deerfield, Illinois received this three-ton IHC Model 61 in 1922. Pirsch called this its "Triple-Combination Pump, Chemical and Hose Car." The body was made of 12-gauge steel. The pump was a "Type B" rated at 300 gpm. The "Champion" 40-gallon steel chemical tank was conveniently placed under the seat. The hose box could hold up to 1,250 feet of 2 1/2-inch hose.

Order No. 2752: This job, on a Packard Twin-Six chassis, was built and delivered to Branford, Connecticut in 1923. Pirsch used mostly Hale pumps, but this triple-combination pumper utilized a Pagel pump rated at 750 gpm. Pirsch could build and supply ladders of either the solid side or trussed design. The iron bar extending in an arch-type configuration across the rear of the hose bed was installed as a "rear safety handhold."

The rugged and dependable FWD chassis, produced in Clintonville, Wisconsin, was requested by a number of communities such as this one from Zion, Illinois. In the Upper Midwest the winters were harsh, which made the rural roads very treacherous. Those were the days before roads were well maintained. But the FWD could go almost anywhere. This triple-combination pumper, built in 1925, contained a type "C" 500-gpm pump. There was room for 1,200 feet of 2 1/2-inch hose and a "Pirsch" 50-gallon chemical tank. The axe holders, side rails, grab rails, etc., were finished in polished brass—not nickel-plated. The lights atop the rear standards were red and green.

Serial No. 558: This picture was taken in front of the 35th Street plant in Kenosha. It is very interesting and warrants some study. With the chassis and pump completed, the rig at this time could undergo its acceptance test. According to Pirsch records, this Model 19, 500-gpm pumper was built for Chaumont, New York in 1926. This was the twelfth "custom" rig sold by Pirsch. Notice the bell and bracket, and the angled toe-board that has the name "Pirsch" inscribed into it. Also notice that the support bracket for the suction hose trough is bolted to the steel frame that will eventually support the driver's seat. From the time the contract was signed to the date of delivery took about 90 calendar days.

Serial No. 568: By the time this city service ladder truck was built for Greenville, Ohio, Pirsch had been producing and marketing its own custom line for about 21 months. It may have been the third ladder truck produced on Pirsch's very own custom chassis. This Model 19, built in 1926, contained a Continental Model 6B motor. The wheels were cast steel discs. Each of the chemical tanks held 35 gallons, but only the ends were nickel-plated. Pirsch's own invention, an electric wire cutter, was among the tools specified in the order.

Serial No. 589: Batavia, Illinois, a far western Chicago suburb at the time, did not have to look very far to find a manufacturer who could build a fire engine. This Model 20 pumper was delivered in August 1927. With its pump rated at 600 gpm, it could pump a full 100 gallons more than the Model 19. Its Waukesha 6KL motor was matched to the larger size pump. The hose body had a capacity for 1,200 feet of 2 1/2-inch hose. The chemical tank was of 40-gallon capacity. Notice the booster reel that is barely visible underneath the tailboard just behind the rear wheel. *Photo Credit: Matt Lee*

Serial No. 596: Washington Township, Ohio liked their first Pirsch pumper so well that they ordered another pumper, identical to the first one, four months later. The one pictured here was the first of the two delivered in December 1927. Power for the 500-gpm pump came from the Continental 6B motor, rated at 70 horsepower. This truck was called a Model 19 "Special." *Photo Credit: Matt Lee*

The Caldwell Township, New Jersey Fire Department purchased this 1928 Day-Elder Model BA6 through the National Motors Manufacturing Company in Irvington, New Jersey. Each of the two "Pirsch" nickel-plated chemical tanks held 35 gallons of water and solution. The acid, when mixed with the soda water solution, resulted in a chemical reaction expelling the contents under pressure through the hose and onto the fire. Pirsch built bodies on only a handful of the Day-Elder chassis.

Serial Numbers 621 (left) & 622 (right): A pair of almost identical Model 19s, shown here, were delivered in 1928 to fire departments in two different states in the same region: Woodland, Oregon (left) and Kalama, Washington (right). Both had the Hale type S 500-gpm pump. Each one had the Continental 6B motor and two 70-gallon copper booster tanks. Both were shipped from the factory on April 14, 1928, on their way to a convention somewhere on the West Coast. Shop orders instructed workers that the trucks were to be "show jobs."

Serial No. 628: The city of Monroe, Wisconsin received two Pirsch custom Model 20 units in 1928—a quadruple combination and this closed cab pumper. Between the two of them, the pumper ended up stealing the show with its cab built by the Stoughton Wagon Works of Stoughton, Wisconsin. It was reported to be the first pumper in the United States with a custom built cab. Due to the care received by the volunteer firefighters, the rig still retains its like-new condition today. It has a 600-gpm pump and two 70-gallon copper booster tanks.

Serial No. 643: Another milestone was reached in 1929 when this Model 16 was delivered to the city of Fond du Lac, Wisconsin. This was the first 1,000-gpm triple built by Pirsch. Pumps of larger capacity required everything else to be more substantial in power and design. Pirsch was able to show their potential for building such a unit to larger cities that required such apparatus. Not shown here was the addition of a large deluge gun that was added later and placed above the hose bed.

Serial No. 749: One of the more dramatic demonstrations of Pirsch fire apparatus came about at the stroke of midnight on April 13, 1931. The fire alarm office relayed an urgent message from Chief Corrigan requesting Pirsch to bring their new smoke ejector machine to Chicago immediately. Twenty-seven men were trapped in a fire involving the sewage system tunnel project far below the ground. Ten firefighters had already perished from the deadly smoke and gases, as did two firefighters that rushed in without masks. The thought was that if Pirsch could bring this machine right away they might be able to save the rest. Peter Pirsch replied that he would send it as quickly as possible. He then called on some of his workers and explained the emergency. In a short time workers began arriving at the plant to assemble the truck. Minneapolis Fire Chief Charles Ringer invented the smoke ejector machine and Pirsch owned the manufacturing rights. The prototype was only in its second stage and sat on the factory floor uncompleted. The men worked through the night to assemble the unit into workable condition. In only a matter of hours it was ready to go! In the wee hours of that morning Pirsch and a couple of his men, along with a rookie policeman who was recruited as its driver, drove the truck and headed towards the tunnel fire, later claiming the 60-mile trip was made in only 88 minutes. The untested machine worked so well that seventeen men were saved. Pirsch's reputation had hit a new high. The rig was eventually purchased by the Hyattsville, Maryland Fire Department in 1933.

Grand Island, Nebraska received this big city service ladder truck in 1931. Built on a White Model 640 chassis, this rig carried 11 ground ladders totaling no less than a whopping 319 feet! The long metal basket atop the double-banked ladder rack carried booster hose, squeegees, plaster hooks, Dietz kerosene torches, and anything else the fire department could fit into the basket. The small booster pump, rated at 100 gpm, received its water from the 100-gallon round booster tank (about one minute's worth of water at full throttle) and was located directly behind the gasoline tank.

THE WORLD'S FIRST ALL-POWERED AERIAL LADDER TRUCK.
BUILT BY PETER PIRSCH & SONS CO.
DELIVERED TO CITY OF SPOKANE, WASHINGTON JUNE 5, 1931.

This is the invention that launched Pirsch's aerial ladder business to new heights. They were now building a revolutionary new aerial that other manufacturers would have to contend with. What originally took two or more men to manually raise "the big stick," was now accomplished by one man. The new hydro-mechanical aerial incorporated at least eight new features that others did not have. The big one, of course, was the ability of one man to raise and lower the ladder, extend and retract the fly sections, as well as rotate the turntable simply by the shifting of levers. What a remarkable achievement! The first hydro-mechanical aerial, an 85-foot, two-piece wooden aerial (Hoist No. 1), was pulled by a White tractor.

P-1502-A

Serial No. 795: This pumper was probably only one of two that did not have a steering wheel. This unusual rig was ordered in 1933 and delivered in 1934 to the U.S. Navy's Lualualei Ammunition Depot on the island of Oahu, Hawaii. This was a Model 21 and could pump 500 gpm. The photograph shows the rig sitting on the rails, which were placed here in a temporary fashion for the photographer. The *real* railroad spur is visible on the right.

P-1560-A

Serial No. 910: The city of Duluth, Minnesota received this Junior aerial ladder truck in 1935. Called a Model 14, it was powered by the Hercules HXE motor. A booster pump supplied water from the 100-gallon water tank. The ladder, which had an electric-powered hoist with a manually operated fly, could extend 55-feet into the air. Pirsch was building three different types of aerials at that time: the Junior model shown here, the Intermediate and the Senior. *Photo Credit: Robert J. Vigliaturo*

THE WORLD'S FIRST 100' ALL POWERED, ALL METAL AERIAL LADDER
DESIGNED AND BUILT BY PETER PIRSCH & SONS CO., AND DELIVERED
TO CITY OF MELROSE, MASS.   JUNE 1936.

Serial No. 933: Pirsch was in the forefront when it came to improvements made in aerial ladder design. Five years earlier they had introduced the first all-powered wooden aerial. However, this time the aerial was of all-metal construction. This Model 14 was delivered to Melrose, Massachusetts. Pirsch had already sold 10 hydro-mechanical powered aerials made of wood, ranging from 65 to 85-feet in length. But this was the world's first 100-foot, all-powered, all-metal aerial made in America.

Serial No. 941: This sporty-looking pumper, a Model 16, was built for Racine, Wisconsin in 1936. The pump was rated at 1000 gpm. The dependable Waukesha 6RB motor supplied power. The J.I. Case Company, a manufacturer of farm machinery in Racine built the cab, which sported a leather top.

This rugged-looking FWD was delivered in 1936 to the city of Ishpeming, Michigan, located on Michigan's Upper Peninsula. In an area known for its heavy snowfalls, this rig was just what was needed to traverse the rough, snow-covered roads. The unit had a 2-stage centrifugal 750-gpm pump.

Serial No. C-37-2: This Diamond-T "Smokeater" quadruple fire engine for Barrington, New Jersey was delivered in 1937. Notice the Dietz hand lanterns mounted on each side of the box. The pump was rated at 500 gpm. The V-type windshield was used on the custom jobs as well. The builder's plate, just below the seat, had two areas for identification: year and type or model. Most of the time only the date of delivery was stamped on the plate.

Occasionally the company performed "metalwork" on trucks for other organizations. The Kenosha chapter of the Red Cross and the local Boy Scout troop assisted the local police and fire department with this 1937 Dodge "humpback" panel wagon specially equipped with all the equipment you see in the foreground. Pirsch did not build the body but did fabricate the tailboard, grab bars, racks for some of the equipment, and mounts for two fire extinguishers. Pirsch also outfitted the trailer.

Serial No. 1084: This sleek-looking open cab quint was built for LaPorte, Indiana in 1939. It is a custom Model 15 chassis with a Waukesha RBR motor. The aerial is a 65-foot three-piece with the pump rated at 500 gpm. The cab itself is a GMC product that Pirsch used on its custom jobs. From the windshield to the front bumper it is all Pirsch metalwork. In the late 1930s and early 1940s, Pirsch was using different grill styles (from wide to narrow and slanted to straight). Note the cutouts in the grill for the bell and combination siren/light.

Serial No. 1110 or 1111: This somewhat "un-American" looking fire engine was delivered to Rio de Janeiro, Brazil, South America in 1940. The curved sheet metal and straight lines give this truck a sleek, streamlined appearance. The midship-mounted rotary pump could pump 500 gpm. The fire hose was in the form of donut rolls and placed in the hose bin on top. The back of the truck contained seats for extra crew members.

Serial No. 1118 or 1119: The city of Chicago, Illinois received two of these nine-man cab sedans in 1940. With the enclosed cabs the crew was afforded protection from the elements and accidents. They were way ahead of their time! The pump was a Hale Model ZMM, 4-stage, 1000-gpm pump powered by a Hercules HXE motor. Water supply was from the hydrant as neither pumper had a booster tank.

Serial No. C-41-09: Simple in appearance yet functional by design, Breckenridge, Minnesota received this open cab GMC in 1941. Very little "extras" were specified here. If a small town did not have an aerial ladder truck, and if some of its buildings were beyond the reach of a standard 24-foot ladder, then occasionally a city service truck was all that was needed to satisfy the insurance requirements of the National Board of Fire Underwriters.

Serial No. 1249: Fire protection at military installations and defense plants was vital. Pirsch built ladders, aerials, pumpers, and yes, trailer pumps as well. Nineteen of these Model 20G pumpers were built for the U.S. Army. They were delivered within a 20-day time period in December 1941. The pump was rated at 750 gpm. Note the absence of chrome—right down to the hose fittings and couplings. This particular truck was earmarked for the U.S. Army's Ordinance Plant at Marion, Illinois.

Serial No. 1201: Memphis was a good customer in 1942. They first ordered this Model 38 that is pictured here and followed-up later with a Senior aerial and two quads. Engine Company 2 (Memphis Fire Department shop no. F-5), with the sedan-style cab, came equipped with a Hall-Scott 243 horsepower 6-cylinder motor. The pumper used a 1,500-gpm Hale 2-stage pump. *Photo Credit: Richard Adelman*

Serial No. 1264: The city of Milwaukee, Wisconsin (Fire Department shop no. 416) received this Model 41 Senior aerial in October 1942. A Waukesha 145GK motor powered the truck. The aluminum ladder was a 100-foot 3-section with hydro-mechanical hoist. Notice the lack of chrome due to wartime restrictions. This truck was eventually resold to the city of Manitowoc, Wisconsin for use as a reserve truck.

Serial No. 1266: This "fireboat on wheels" was actually built and developed by the Memphis Fire Department shops along with a local firm. Originally a quad, it was one of two Pirsch Model 27s delivered in 1942 (similar to the one shown in the photo inset). When new, it featured a 750-gpm pump with a 100-gallon booster tank. It was around 1954 when the quad underwent a radical change to a deluge rig. The pump size was increased to 1,000 gpm and a 500-gallon tank was installed. The deck gun with its 3-inch tip could discharge 3,000 gpm! The rig was equipped with wet water, which enabled it to penetrate deep-seated fires at lumberyards and large warehouses used to store large quantities of combustible material. It was given Memphis Fire Department shop no. F-359. *Photo Credit: Richard Adelman*

Serial No. PB-46: This compact, yet busy-looking little squad unit, was built for Waukegan, Illinois in 1944. The chassis is an IHC Model K-5, which was introduced in 1940 and built at IHC's Springfield, Ohio plant. The Hale Model H booster pump was rated at 100 gpm and supplied from a small booster tank. A large number of fire departments utilized squads of various sizes and styles to bring extra personnel and equipment to fires and other emergencies.

Serial No. 1459: A Sterling Model 20 Sirenlite on the nose cleared the way to the fires for this 1945 custom pumper, which was also a Model 20. Don't let the name fool you. This rig went to a small town of Oakland, Iowa. The suction inlet and discharge gates for the 500-gpm pump are located behind the door, just below the gauges. Access to the bench seat under the canopy was through a center aisle extending from the rear of the hose bed to just under the canopy.

Serial Numbers 1590 through 1593: Four identical Model 38 custom pumpers were delivered on the same day in October 1946 to Tacoma, Washington. These 1,000-gpm engines all had GMC closed cabs. Note that the headlights have painted housings but the bumper, siren and spotlight are chrome. Wartime restrictions were starting to dwindle. The intake (suction) manifold on the side is siamesed with the addition of two 2 1/2-inch inlets; a feature not found on very many trucks.

Serial No. 1594: Somewhat unusual looking was this Model 20, built in 1947 with a 55-foot 3-piece Junior aerial ladder. From the front grill to the windshield the steep incline of the hood is quite noticeable. Bridgeton, New Jersey received the 39th Junior aerial made. Pirsch made the wood trussed ladders that you see in the racks.

Serial No. PC-21: Pirsch sales were very strong in the southern United States during the late 1940s era. Sharing in the deliveries was this stubby-looking Ford COE built for Lakeland, Florida in 1947. The short wheelbase would allow the driver to maneuver his rig almost anywhere. This pumper came equipped with a Model ZL 500-gpm pump made by Hale. The total sales price (minus the chassis) was only $6,350.

Serial No. PC-57: John Bean Fire Apparatus Division of the FMC Corporation, was not the only manufacturer of high-pressure pumps (but probably the most famous). The town of St. Albans, Vermont, a faithful Pirsch customer, purchased this 1948 IHC Model KB-7. The Hale Model HPZL combination high pressure/volume pump would supply 75 gpm at 600 psi or 500 gpm at 120 psi. With 500 gallons on board, the pumper was a versatile firefighting unit. The three large floodlights provided ample lighting for nighttime fires.

Serial No. 1770: This Pirsch Model 41E Senior 100-foot 3-piece aerial was delivered to Windsor, Ontario, Canada in April 1949. The hoist (serial no. H-269) is hydro-mechanical. The ladder was raised by hydraulics while the turntable rotation and the extension of the fly sections were moved mechanically. Pirsch cabs were the norm on most custom chassis but they would on occasion supply the customer with a GMC cab as shown here.

Serial No. PC-61: In Minnesota, where you'll find some of the coldest temperatures in the U.S.A., Bemidji opted for a 1949 Chevy pumper with its cab-top removed! However, the smaller appliances as well as the pump controls were well protected behind the compartment doors. The unit had a Hale "ZL" 2-stage 500-gpm pump. Pirsch bodywork is evident from the windshield back to the tailboard. Note the large, square box above the hose bed.

Serial No. 1798: Mattoon, Illinois received this Model 60D "intermediate" aerial ladder in 1949. It was equipped with a 750-gpm pump and a 65-foot 3-piece aerial as well as a small booster tank. Note the tank fill line with a shut-off valve located just below the hard suction. The word "intermediate" implied that the ladder's turntable was center-mounted on the truck. The Waukesha 145GZ motor supplied power for both the pump and aerial.

A couple different variations of quadruple combinations are shown here. The job for Burlington, Wisconsin (top), Serial No. 1815, Model 64A, had a 750-gpm pump, two lengths of suction hose mounted on the side, and a hose box above the ladders as well as a hose trough behind each set of rear wheels. It was delivered in 1949. The truck shown below, also delivered in 1949, has no cab doors, one length of hard suction on the side, compartments running almost the full length of the body, and one hose reel located at the rear under the ladders. The Gibbstown, New Jersey quad is a Model 20, Serial No. 1796, with a 500-gpm pump. *Photo Credit for photo shown below: Kerry Poltrock*

Serial No. 1873: The Pirsch Company built a special rescue/pumper in 1949 for the fire department that protected them. This large rig also sported a 750-gpm pump concealed behind a compartment door just behind the cab. Called a Model 64A, the four-door cab provided enough seats for a large crew of firefighters. It was Kenosha's first rig in which firefighters did not have to ride on the tailboard.

Serial No. PC-77: Protection of rural areas that had no hydrants was the forethought here in Somers, Wisconsin in 1950. This large tanker, built on an IHC chassis, hauled 1,850 gallons of water but could supply 250 gpm with its own booster pump at the fire scene if needed. With all of the extra equipment carried it was virtually its own fire department on wheels! This may have been the first large round tanker of its type built by Pirsch, but it would not be their last.

Serial No. 1976: Pirsch's customers were not just municipal fire departments. Industrial giants such as the Dow Chemical Company at Midland, Michigan were large enough to have their own fire department. The pumper shown here was delivered in March 1952. It was a Model 41A and had a 750-gpm pump. The IHC "Space-Saver" cab could be utilized on Pirsch custom chassis, too. Note how the sheet metal has been used to integrate the nose with the much wider cab by streamlining the sections together.

Serial Numbers PE-07 or PE-08: In 1953, the fire department of DeKalb County, Georgia ordered and received a pair of open cab GMC pumpers with 750-gpm pumps. Similar to the Bemidji rig shown on page 54, this rig also uses Pirsch sheetmetal in streamlined fashion from the cowl on back. Note the small Pirsch nameplate barely visible located on the bottom of the rear fender. The larger nameplate, like the one shown on the next page, was not used on commercial jobs.

Serial No. 2140: The city of Sarasota, Florida received this Model 64BZ intermediate aerial truck in 1954. The 85-footer, shown here, consisted of four sections. The control pedestal when mounted on the driver's side meant that the hoist was a hydro-mechanical one. If mounted on the opposite side (officer's side), the aerial was powered by a full hydraulic hoist. Waukesha Motors, located in Waukesha, Wisconsin, was the supplier for a majority of the motors that provided the power for Pirsch's own custom line of fire trucks.

Serial No. 2216: This rig with squared fenders was delivered to New Boston, Ohio on September 30, 1955. The pump was rated at 1,000 gpm and was powered by the popular 145GKB Waukesha motor. The squared front fenders were the first major change in decades, as the round fenders would be eventually phased out in time. This canopy cab style found favor with a large number of fire departments. It provided some protection for the crew riding in the rear jump seat behind the window that you see above the pump panel. Access was either from the back or from one of the two sides of the truck.

Serial No. PE-58: Since the early 1950s Pirsch had been building more and more bodies with the squared rear fenders, with or without compartments. Little by little, firefighting equipment was disappearing off the running boards, stashed away inside and out of the elements. This job was built for Lumber City, Georgia in 1955 on a Chevrolet 6400 model. The pump was rated at 500 gpm. The lack of chrome, only one combination siren/light, and simple design made this a very economical pumper at only $4,950 not including the chassis.

Serial No. 2261: Dallas, Texas received this Senior 100-foot 3-piece aerial in June 1956. Pirsch called this their Model 41E because of its Waukesha motor and tillered aerial. This was the fourth Senior aerial delivered that year. Total cost was $38,296. Notice the huge cord reel located just behind the turntable. This style of cab was called the "open driver's compartment." The tillerman's seat would swing up and out of the way very quickly when the aerial had to be raised—another Pirsch exclusive.

Serial No. 2267: Minneapolis received this rescue squad in 1956. A 300 horsepower Hall-Scott motor powered the Model 80. The Gerstenslager Company of Wooster, Ohio fabricated a major portion of it, from the windshield back to the tailboard. Once in awhile Pirsch would outsource the bodywork, and rescues were no exception. This "one-of-a-kind" rig would sadly end up being demolished in an accident many years later. *Photo by Steve Skaar/Paul Barrett Collection*

Serial No. 2298: Fire engines built for the city of Baltimore, Maryland Fire Department were easily recognized by their distinctive red and white paint schemes. This Model 41B was delivered in December 1956. The motor was a Waukesha 145GKB. The pump was rated at 1,000 gpm. Squared front fenders introduced only a few years earlier were now gaining popularity. Note the open compartment area behind the rear wheel. This was built to Baltimore's specifications.

Serial Numbers PF-58 or PF-59: The F-series Ford truck chassis were very popular and provided the customer a wide range of different sizes and options. No wonder then that at least 25 percent of the commercial jobs were on the F-series. DeKalb County, Georgia received two pumpers on the F-850 model in 1958. The pumps were rated at 750 gpm. Total costs for the bodywork and pump for each truck was $14,393.

Serial No. PF-87: The FWD Corporation of Clintonville, Wisconsin provided just the cab and chassis for this big ladder truck in 1958 for Syosset, New York. The aerial shown here is a 75-foot 4-piece of all-aluminum construction. During the course of the late 1950s and on into the early 1970s, the FWD fire truck was a popular choice and sales were very strong. However, probably less than 25 made their way into the plant for customers who desired Pirsch bodywork or aerials mounted on FWD chassis. Pirsch aerials were mounted on a variety of commercial and custom chassis such as Maxim, Mack, Hahn, American LaFrance and Ward LaFrance. *Photo Credit: Larry Phillips*

Serial No. 2479: This Senior 85-foot 3-piece aerial ladder sported a few extras, from the chrome bell on the front fender to the deck pipe you see located behind the steering wheels on the trailer. The all-aluminum aerials were popular in many large cities across the United States, including this Model 41E built for Rochester, New York in 1960. The two-tone paint job makes this a nice-looking truck. *Photo Credit: Matt Lee*

Serial No. PG-79: Pirsch delivered this GMC Model V5008 to Hubbard, Oregon in 1960. At first glance this appears to be just another commercial job. Look closely and you will see vents protruding out in horizontal fashion on each compartment door allowing the insides to "breathe." This helped to dissipate any moisture and prolong the life of the bodywork. The pump was a Hale Model 2QLD75-2, rated at 750 gpm. *Photo Credit: Carl Groth*

Serial No. PH-03: This rescue squad on a Ford F600 chassis was built for Bloomingdale, Illinois and delivered in 1961. Squads built by Pirsch were not all that common. They were mostly found in fire departments that already had Pirsch fire apparatus. This squad had a Hale CBP-5 booster pump that produced 200 gpm at 150 psi. Booster tank capacity was minimal at just 150 gallons. *Photo Credit: Carl Groth*

Serial No. PH-32: Louisville, Kentucky was a good customer of pumpers, aerial trucks and even this chemical unit built by Pirsch. Mounted on a 1962 Ford C-1000 chassis, this rig was built and designed to handle a wide variety of fires that could not be extinguished with plain water. The truck contained ramps for unloading two portable chemical carts that were stored in open compartments at the rear of the truck. They could be pulled by hand to a fire that required dry chemicals for extinguishing. Other agents such as AFFF, protein foam, and carbon dioxide were also carried on board. The pump and booster tank sizes were each at 500-gallon capacities.

Serial No. PH-82: The Sutphen Fire Equipment Company located in Amlin, Ohio at the time, was a sales agent and distributor for Pirsch. Customers had their choice between buying Sutphen's own line of fire apparatus or buying a Pirsch, depending on the type and style of fire truck that was required. Pirsch first sold the rig (you see pictured here) to Sutphen as a "demonstrator" unit. In time, Sutphen sold this IHC CO-8190 model to Elkins, West Virginia later in 1963. The pump was rated at 750 gpm.

Serial No. 2695: This big 85-foot elevating platform manufactured by the Pitman Snorkel Company was the fourth one of various heights that Pirsch bodywork was done on. Rehoboth Beach, Delaware received this custom Model 41S in 1964. At $55,615, the price of this rig was a full $10,000 to $12,000 higher than that of a custom Senior 100-foot aerial! Notice the wheel cutouts at the rear and how they flatten out just above the wheel. This allowed fire equipment to be mounted at a slightly lower height, making retrieval much easier.

Serial No. 2696: This 1,000-gpm pumper Model 80 was delivered in January 1964 to Portland, Oregon. The Model 80s had a Hall-Scott motor. Once a popular motor, fewer Hall-Scott's were being ordered by this time. Waukesha motors were now the preferred choice by the majority of customers.

Serial No. PI-48: The popularity of the elevating platform was increasing more and more. During 1964, Pirsch fabricated the bodywork on this Ford C-950 for Vandergrift, Pennsylvania. The pump was rated at 750 gpm. The 50-foot boom was Pitman Snorkel's shortest model. Pirsch's sales price for the platform, pump and bodywork (excluding the chassis), was a mere $31,235. It was just one of six Snorkels delivered that year. The year of 1964 was a busy year with no less than 95 trucks having been built and delivered.

Serial No. PI-53: This rig has probably sparked more than a few arguments among those who claimed it was a custom versus those who said it was a commercial. This is one of those cases where the line between the two gets somewhat fuzzy. Underneath the custom cab lies a stretched Ford C-model chassis. This rig was given a commercial serial number because the chassis was *not* a Pirsch product. This is another reason why it is *very* important to note the serial numbers of the body and chassis builders. They not only provide positive identification but can assist the historian/photographer in many other ways as well. This rig was built in 1964 for Middletown, Kentucky. It has a 1,000-gpm pump and a 750-gallon tank. *Photo Credit: Mark Stampfl*

Serial No. 2711: Quite a marriage this was for McLean, Virginia in 1964. The Gerstenslager Company of Wooster, Ohio built the body shown here. The chassis was a Model 41R ("R" stood for rescue). A Hale CBP-5 model, 250-gpm PTO pump with a small booster tank was also included. Note the Buck-Eye Roto-ray on the cab's roof.

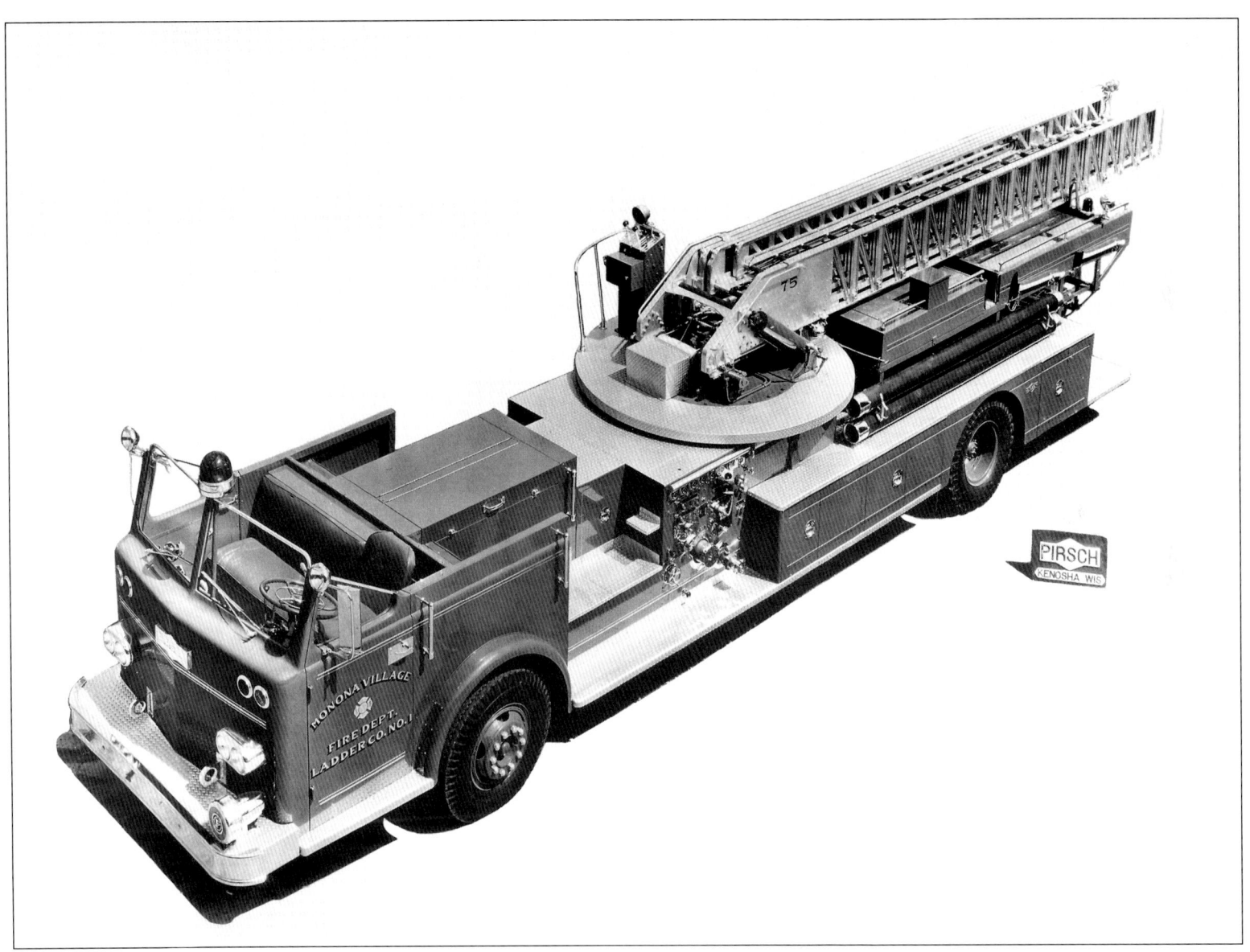

Serial No. 2763: This Model 41BQ was built for Monona, Wisconsin in 1965. It has a 1,000-gpm pump, 250-gallon water tank, and a 75-foot 4-piece intermediate aerial. The Pirsch custom cab provided spacious seating for five firefighters. The covers on the hood opened in "butterfly" fashion, permitting easy access for servicing the motor. This rig is still in service today and is in almost "like new" condition.

Serial No. PK-03: In 1967, Reading, Pennsylvania received this one-off open cab, 75-foot elevating platform on the IHC Model CO-8190 chassis. Notice that the top center of the windshield has been notched out to allow the lower traveling height. With no cab roof it was probably easier to fit in the station. The Snorkel Fire Equipment Company of St. Joseph, Missouri manufactured the platform. Notice the pompier ladder just above the extension ladders. Pirsch's sales price, minus the chassis, was $37,348.

Serial No. 2813: In addition to this Senior aerial Jackson, Tennessee also received two pumpers in 1967. The 100-foot 3-piece shown here was pulled by a custom cab-forward tractor of standard design. The trailer had full compartmentation along the lower half. The outriggers were manually operated. The tiller seat was of the Pirsch quick swing-away type. In the smaller photograph, note the saddle that was designed to provide nesting room for the ladder pipe attached to the base section of the aerial. Having the ladder pipe pre-mounted saved precious time when it was needed.

Serial No. 2827 or 2828: Cedar Rapids, Iowa was another strong Pirsch customer for many years. In March 1967 a pair of Model 41Cs were added to the fleet. Although this looks like another cab-forward pumper, there are a few features here that we can pick out as being slightly different. The crosslays that you see here were not all that common yet. The warning light package included a pair of "lollypop" flashers on the roof. These types of flashers were very popular on fire trucks produced by General Fire & Safety of North Branch, Minnesota.

Serial No. 2874: Occasionally, a semi cab-forward pumper rolled off the line at the plant. This pumper built for Chevy Chase, Maryland in 1968 had all the usual characteristics including "D-ring" door handles and those protruding rear fender crowns. This style of fender was uncommon except for some of the deliveries in the Maryland area. The pump was rated at 750 gpm but there appears to be more than three discharges judging by the number of levers with round, black handles. The motor was the standard Waukesha Model F817G. Note the storage area for the soft suction hose cut into the cab face just below the Federal "Q."

Serial No. 2903: This black and white picture falls short of showing the true beauty of this 1969 custom pumper Model 41B built for the Rouss Fire Company in Winchester, Virginia. Its companion, a Pirsch ladder truck, was done up in the same color and style. The pump was rated at 1,000 gpm with a 400-gallon booster tank. *Photo Credit: Gene Brotzman*

Serial No. 2923: Poised and ready for action, this 1969 custom Pirsch Model 41B sits on the apron in front of the Macedonia, Ohio fire station. A Waukesha F817G motor provides the power for the 1,000-gpm Hale pump. The "Mr. Ed"-type door (partial door) just behind the pump panel and above the diamond-plate provides access to the crew cab area at the rear of the cab. Another way to access this area was found on some of the earlier models that had a walkway through the center of the hose bed. *Photo Credit: John Sytsma*

Serial No. PK-85 or PK-86 or PK-87: Cleveland, Ohio received three 100-foot 3-piece Senior aerials of this type in September 1969. The tractor was a Ford C-1000 model. Tractor-drawn aerial ladders, whether they were pulled by a custom Pirsch or a commercial tractor, were still referred to as the "Senior" model.

Serial No. 2905: Skokie, Illinois was another one of those Chicago suburbs that had a large number of Pirsch fire trucks in its fleet. This Model 41S with a 65-foot elevating platform built by Snorkel was delivered in 1969. Power was provided by the Waukesha F817G gas motor. A number of fire trucks throughout the Chicago area sported the type of lights shown here—Mars DX-40 model lights on top of the cab. *Photo Credit: Gary Kadzielawski. Charles Madderom Collection*

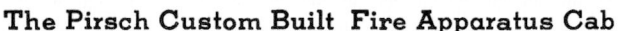

The Pirsch Custom Built Fire Apparatus Cab

Pirsch offers a variety of fire apparatus cabs for their trucks which are custom built in the Pirsch plant of the best materials by craftsmen having years of experience in metal fabrication. Heavy duty bracing provides extra protection to occupants in the event of an accident.

The Pirsch built cab is all metal construction with 10 gauge steel frame, 16 gauge dash and 16 gauge cold rolled stretcher levelled steel roof, doors and panels. There are no wood fillers preventing warpage and expansion caused by wood. The men riding inside of the cab are provided with exceptional protection from severe weather conditions. Windshields and windows are of safety plate glass, of generous size to provide wide visability.

The instrument panel, light and ignition switches, etc., are all conveniently located on the dash and a glove compartment with door is located on the right hand side of cowl. The wiring is vari-colored and connections properly labelled to provide ease of checking. Panel includes sight gauges. Dome lights furnished in cab and in canopy.

The inside width of the Pirsch cab provides comfortable seating room for the crew. The inside height provides maximum head room and increased windshied visability which is of such great importance in todays congested traffic conditions. Durable long lasting upholstery is provided.

Since wood fillers can cause warpage and corrosion, they have been completely eliminated.

The type of cab offered in this bid is shown on "Data Sheet."

CAB FORWARD TYPE

ENGINE FORWARD TYPE

PIRSCH

Both cab styles shown here were custom built at the Kenosha plant. The "engine forward" type was introduced in 1955 while the "cab-forward" type (shown here in its 1970s style) made its debut six years later in 1961.

Serial No. 2961: Cab-forwards were favored over conventionals by a wide margin, but that didn't stop Evendale, Ohio from receiving this Model 41C in 1970. Power was supplied to the 1,250-gpm pump by a Waukesha Model F817G motor. This sharp-looking pumper has an ample assortment of warning lights and sirens. Notice the "tunnel lights" built into the cab's roof. This was a Pirsch signature, used on the style of cab shown here as well as the cab-forward units. *Photo Credit: John Sytsma*

Serial No. 2982: College Park, Maryland was the customer for this unusual low-profile aerial built in 1971. The 100-foot 4-piece shown here was referred to as the "intermediate" model, even though it was rear-mounted. Although it had a Detroit 8V-71N Diesel motor it was still called a Model 41DT. Later on, custom jobs having a Detroit Diesel motor were given a model number of 86 or 88; the last digit referred to the number of cylinders. This was probably the only low-profile aerial of this style ever built.

Serial No. PL-34: Akron, Ohio ordered this 85-foot 4-piece intermediate aerial in 1970, and received it in 1972. This was an indication that orders were coming in faster than the units were being produced. One look at this rig would give you the impression that this is a custom Pirsch, but the two small square mesh grills just below the headlights show this to be a Duplex chassis with a TCM (Cincinnati) cab. Pirsch did not start using the TCM cabs in any quantity on their own custom chassis until around 1975. Note that on the cab's face a small Pirsch nameplate has been installed in place of the larger, more customary nameplate.

Serial No. PL-55: Roberts Park, Illinois received this nice-looking Mack CF-685 quint in 1972. Pirsch produced a number of jobs such as this one on the Mack C and CF chassis. The rear-mounts, and the intermediate style shown here, were popular choices for fire departments, especially in the Chicago area. The aerial was a 100-foot 4-piece aluminum ladder. Waterous pumps were usually found on the Mack CF chassis but this one had a 1,000-gpm Hale pump. Pirsch used the Hale pump on almost all their jobs.

Serial No. PL-72: Fire trucks built on International's Fleetstar model were scarce at Pirsch. This chassis had a much higher GVW rating than its smaller companion, the Loadstar. The Thompson Grove Fire Department located in Cottage Grove, Minnesota purchased this IHC 2000-D model in 1972. It had a 1,000-gpm pump. *Photo Credit: Paul Barrett*

Serial No. 3063: Memphis, Tennessee received a pair of Model 88D ladder trucks in 1973. Each of them had a 75-foot 4-piece intermediate aerial along with a 250-gpm booster pump and small tank. The motor was a Detroit 8V-71N Diesel. Note the pompier ladder carried on the side. Pictured is Truck 11 (Fire Department shop no. F-127), taken at the Memphis Fire Department Training Academy. *Photo Credit: Richard Adelman*

Serial No. PL-76: Not all Fords have to look the same! The McMahan Rural Fire Protection District near Louisville, Kentucky received this longer-than-normal pumper on the C-900 chassis in 1973. It was originally set up as a quadruple combination. Quads at one time were more common, but occasionally a fire department ordered one during the 1960s and 1970s. The pump was rated at 1,000 gpm while the tank held 500 gallons. Note that the letters F-O-R-D have been removed from the front face of the cab and replaced by the large Pirsch nameplate.

Serial No. PM-07: The first delivery made in 1974 was this Senior 100-foot 4-piece aerial pulled by a Mack CF795FC model tractor headed for Warren, Michigan. Those outriggers that you see tucked under the turntable in their traveling position stabilized the truck and aerial when they were extended out and then down. Unlike today's hydraulic operated jacks, these were manually operated. Notice the enclosed cab for the tillerman. The ones with the sliding doors, such as this one, were actually J.I. Case Company tractor cabs!

Serial Numbers 3088 (left) and 3087 (right): Fire trucks built for Louisville, Kentucky were easy to distinguish because of the diamond tread that covered the bottom portion of the pump panel and compartments. Shown here are a pair of Model 88B pumpers delivered in 1974. Both featured a 1,000-gpm pump, a 500-gallon tank and a Detroit 8V-71N Diesel motor.

Serial No. PM-31: Orders for the shorter Tele-Squrts were more numerous than for the 75-footer you see here. This big truck was received by Oak Creek, Wisconsin in 1975, a full two years after the bid was awarded to the Carl Groth Sales Company. The chassis is a Ford Q-8000. The pump is rated at 1,000 gpm and the booster tank holds 300 gallons. Carl Groth, a Pirsch Company "rep" (salesman) had the Pirsch dealership in Wisconsin and owned the franchise for aerials and water towers made by the Snorkel Fire Equipment Company located in St. Joseph, Missouri. Total selling price was $82,000. *Photo Credit: Carl Groth*

Serial No. PM-55: Delivered to Waycross, Georgia in 1975, this IHC Model CO-1910B has a Hale Model QLF125-4 pump. Pirsch made only a few bodies on this style of IHC chassis. This rig has a wrap-around grab rail at the rear of its canopy cab, a feature found on many rigs having a tilt-cab canopy-style cab. The long chrome cylinder above the gauges contained small lights that illuminated the pump panel for nighttime operations.

Serial No. 3130: Montgomery, Ohio received this Model 88C in 1975. Tunnel lights on the roof, extra pin-striping, and chromed front fenders gave this pumper a snazzy look. This rig could pump 1,250 gpm. The booster tank held 750 gallons. Note the absence of hard suction hose and the round holes cut into the diamond-plated doghouse to facilitate additional cooling for the Detroit 8-cylinder Diesel motor.

100

Serial No. 3166: This is thought to have been the last conventional-style custom truck built by Pirsch. It was delivered in 1976 to the Springfield Fire Department in Summit County, Ohio. This Model 86B had a 1,000-gpm pump and a 750-gallon booster tank. Under the long-nosed hood, a 6-cylinder Detroit Diesel provided the power. Look closely and you will see the hinges and door handle on the side just in front of the suction hose. This was the "Mr. Ed" door that was used by the crew for gaining access to the crew cab area behind the front seats. To this day, the location of this rig remains a mystery. Information regarding its status would be most welcomed by the authors.

Serial No. 3192: Evergreen, Colorado received this lime-green colored Model 88C pumper in September 1977. The scales were now tipping in favor of the TCM (Cincinnati) cabs. Pre-made cabs made it possible to shorten the number of man-hours needed to "pound out" a rig, thus making it more attractive to customers wanting a shorter delivery time, which translated into additional sales. The pump was rated at 1,250 gpm with a 750-gallon booster tank.

Serial No. PN-19: Read the lettering on the side. Were pumpers like this the forerunners of today's big combination pumper/ rescue trucks? If so, the number of compartments has really increased! This rig was built in 1977 for Shamokin, Pennsylvania. Despite the appearance as a Pirsch "custom," the large chrome belt wrapped around the front face of the cab gives this away as a Hendrickson. This was their Model 1871C chassis. The top-mount pump controls we find so common today were just beginning to take hold during this era. The pump was rated at 1,000 gpm and the booster tank held 500 gallons.

Serial No. 3202: Lisle FPD serving the communities of Lisle and Woodridge, Illinois received a pair of Pirsch custom pumpers in 1978. This one sported top-mount controls while the other had the controls in the traditional spot on the side. This one is a Model 88C. Note the large array of warning lights and siren here, from the two Mars Borealis lights on top to the Federal "Q2" siren recessed in the front face. "High-side" compartments and the small "quarter compartments" for air bottles shows this rig to be ahead of its time. Controls on top gave the pump operator a maximum view of the fire ground area surrounding the truck. The Hale Model QSMF125 was rated at 1,250 gpm and the booster tank was larger than the normal size for this time period at 750 gallons.

Serial No. PN-81: Although not as common as the Ford C-models, this picture shows that Pirsch would build on just about any chassis that was brought in for a "metal-bending" job by a customer. During the entire decade spanning the 1970s, orders on this Chevrolet 60 series chassis totaled less than a "handful." This Chevrolet COE was delivered to Glasgow, Kentucky in 1978. The pump was rated at 1,000 gpm. The chrome ring with a mesh insert is the screen that hides the radio speaker and protects it from the elements.

Serial No. 3226: About 1972, Pirsch started the practice of exposing the cab steps by incorporating cab doors of 3/4-length on their custom cab-forward chassis. One exception was this 1979 Model 86C pumper delivered to Ironton, Ohio with full-length doors. It had a Hale QLF 125-2 pump rated at 1,250 gpm, with a 300-gallon tank. Note the gabled cover on the "doghouse" covering the motor. Most generally, but not always, this was an indication that the power plant was a 6-cylinder instead of the larger 8-cylinder.

Serial No. 3206: The city of Evanston, Illinois eventually took delivery of this Model 88C custom pumper in January 1980. This was Pirsch's "Stock Model D." In an effort to pick up additional sales, stock jobs were produced right alongside other rigs with signed orders. A variety of them (pumpers and aerials) were produced over the years. The prospective customer could buy a stock job, with little or no modifications done, and have it in a fraction of the normal delivery time. It featured a 1,250-gpm pump and the tank held 750 gallons. The motor was a Detroit 8-cylinder.

Serial No. PO-33 (Stock No. 37): A majority of the commercial orders were built on the Ford C-series chassis throughout the late 1970s and early 1980s. This C-8000 model, built for Starkville, Mississippi in 1981, had a 1,000-gpm pump and 500 gallons of water. The crosslay box under the crew cab was a space-saving idea, designed by Pirsch engineer/draftsman Bill Kaeppeler (pictured in photo inset). Fire chiefs liked what they saw. The "Kaeppeler Crosslay" was utilized in a number of rigs with the Ford C-series chassis. *Photo Credit: Carl Groth*

Serial No. PO-55: This combination hose and manifold wagon delivered in 1981 is a very interesting piece. Fighting fires was done with a different concept in Auburn, Alabama. This is a Ford LN-8000 with a CAT 3208 motor and an Allison MT643 transmission. The reel was large enough to hold 1,000 feet of five-inch LDH hose.

Serial No. PO-61: Chanhassen, Minnesota received this LS-9000 Ford chassis with a 1,250-gpm pump in 1981. This style of Ford chassis, the LS series, was very popular in Minnesota—more notably in the Twin Cities area. The 4-door "commercial" cabs of this type provided the crew with excellent protection from the harsh winters at a cost lower than that of a "custom" built cab. Notice that the body is actually constructed in two separate pieces.

Serial No. 3282: Late in 1981, Pirsch ceased production of its own custom cab and started using the TCM (Cincinnati) cab in its place. If a customer wanted a Pirsch custom cab and chassis the TCM cab, built in Cincinnati, Ohio, is what was used. Pirsch custom cabs were truly works of art built by skilled craftsmen. In their efforts to remain competitive a more cost-effective alternative was to purchase cabs that were mass-produced. This 100-foot 4-piece, rear-mount ladder truck, Model CAN8 with a Detroit 8V71N motor, was delivered to College Park, Georgia in 1982. *Photo Credit: George Reichhardt*

Serial No: PO-64: There are a number of volunteer fire departments in the Midwest with class and distinction, but Mount Horeb, Wisconsin ranks near the top. This picture shows why. Plenty of chrome, pin striping, and bells and whistles adorn this good-looking 100-foot rear-mount quint that was delivered in early 1982. Built on the Spartan chassis, it was powered by a Detroit Diesel motor. The pump is a 1,250-gpm Waterous with a 200-gallon booster tank. With the exception of an older American LaFrance pumper, all of the department's front-line trucks are now painted black. This rig is now servicing Litchfield, Illinois. *Photo Credit: Carl Groth*

Serial Numbers PO-78, PO-79 or PO-80: The Pirsch aluminum ladder when used and cared for properly would outlast most chassis and bodies. A large number of aerials, such as the one shown here for Memphis, Tennessee in 1982, were rehabbed throughout the years. This Senior aerial, built in 1937, was originally pulled by a Pirsch custom tractor. Memphis decided that since the ladder was in solid condition and pulled by a new tractor, it made economic sense to rehab it. This trailer when new had round fenders. The trailer has been completely refurbed with new sheet metal and compartments. Memphis was an excellent Pirsch customer over the years, purchasing no less than 97 units. This number also includes the ones that were rebuilt.

Serial Numbers PO-83 through PO-99: In one of the largest municipal contracts ever awarded, Pirsch won the bid to supply the city of Houston, Texas with an order of 16 pumpers on Spartan chassis. They were all delivered in a nine-month period between 1982 and 1983. Each pumper sported a white paint job, a 1,500-gpm pump and a 500-gallon booster tank. A Detroit Diesel 8V-71N motor, coupled with an Allison HT740 transmission, provided the power. The cost for each one was $138,000. Shown is E51, which was assigned Serial No. PO-93 and Pirsch Job number 10.

Serial No. PQ-28: Center Moriches, New York wanted their pumper body constructed of aluminum. Any order that specified a body to be of aluminum construction was sent out to The American Modular Corporation, a subcontractor for Pirsch, located in Smithfield, Rhode Island. The unpainted bodies were then sent back to Pirsch for painting and final assembly. The pumper shown here is a 1983 Ford C-8000 model with a 1,250-gpm pump and a 750-gallon tank.

Serial No. PQ-44: Opelika, Alabama received this low-profile Hendrickson Model 1871CS early in 1984. It had a Detroit Diesel 6V92TA motor and an Allison MT643 transmission. The pump was rated at 1,250 gpm. With the Pirsch nameplate on the front, this appears to be a custom Pirsch, but it is not. On many occasions commercial nameplates were eliminated in favor of the Pirsch nameplate. The only true way to determine the difference between a commercial or a custom job was to look for the serial number stamped on metal plates located in and/or on the truck. Customs were generally assigned a 3- or 4-digit number beginning in 1924. The longest running commercial list began with two letters followed by two numbers beginning in January 1942.

Serial No. 3307 (Stock Model P): One of the first "Skytop 110" models to be produced went to Cleveland Heights, Ohio in April 1984. This quint, a Model CFC8, had a 1,250-gpm pump and a 200-gallon booster tank. This new concept and design was the first major change in Pirsch's famous lattice design in decades. The company was given the Governor's "New Product Award" for their contribution to the economy of Wisconsin. The award was presented based on the concept, function, safety, and appearance of the design.

Serial No. 3334: This job, built for and in the city where Pirsch was located, sadly would be the last one ordered by the Kenosha Fire Department before the plant's closing in the late 1980s. The 1984 Model CPC6 was powered by a Detroit 671N Diesel coupled with an Allison MT644 automatic transmission. The pump was a Hale 1,250-gpm. It carried 500 gallons in its booster tank. The first two letters of the model number stood for Custom Pumper. The third letter "C" meant the pump was of 1,250-gpm capacity. The numeral "6" meant the motor was a 6-cylinder.

Serial No. 3349: Shown against a blue sky on the shores of Lake Michigan, not far from the 35th Street plant, is this 1984 custom Senior 100-foot aerial Model CSN7 built for Buffalo, New York. The power plant is a Cummins LTA10 and it has an Allison HT 740D transmission. This job was one of several pumpers and aerials built by Pirsch to replace the rigs that were damaged in a previous explosion and fire in Buffalo. The aerials for Buffalo were some of the last ones built using the older-style lattice design. A new aluminum aerial of different design had already made its debut.

Serial Numbers 3353 and 3354: Houston, Texas only a few years earlier had taken delivery of 16 pumpers built on the Spartan chassis. E-17, shown here, was one of two that was ordered on the Pirsch Model CPD8 custom chassis and delivered late in 1984. The pump was rated at 1,500 gpm. The booster tank held 500 gallons of water. Unlike the previous order, where the pumpers sported a white paint job, these rigs were painted all red. Look closely and you will notice the dark background on the Pirsch nameplate. The deep blue color that contrasted with the chrome letters was chosen to better reflect the company's name. This minute change was employed on some, but not all jobs, starting about 1982.

Serial No. 3366: Barranquilla, Columbia would be the last city anywhere in South America to receive a Pirsch fire truck. The 1985 Model CFC6 pictured here was built as another stock model. The pump was rated at 1,250 gpm and the tank held 500 gallons. This 110-foot ladder was the eighth "Skytop" produced. Note the wooden crating on top of the rig. Workers were making preparations to secure extra equipment for its long journey south. It was driven under its own power to a southern seaport and was then loaded aboard a ship.

Serial No. PQ-53: Fern Creek, Kentucky received this behemoth-looking tanker/pumper on a White chassis in 1985. The pump was rated at 1,250 gpm and the water tank held 2,000 gallons. It would seem this rig almost had to be "pushed" in order to get it through the plant's main door on the 35th Street side of the building. Other than the awning overhead the door on the left, the appearance of the front of the plant remained virtually unchanged since it was built in 1916.

Serial No. 0028: This tanker/pumper combination was purchased by Chaska, Minnesota in 1988. It was delivered during the Hass/Blondek era. This was the reason for the new serial number given. The pump was rated at 1,500 gpm and the tank at 2,500 gallons. *Photo Credit: Paul Barrett*

Serial Numbers 0040, 0043, 0044 and 0045: Four of Chicago's older Pirsch aerials, originally placed on 1966 Mack C-85 chassis, were scheduled for rehabbing on 1985 Ford C-8000 chassis when the plant ceased production in September 1986. Eventually, under the ownership of Haas/Blondek, Inc., Pirsch completed the task of rebuilding the aerials and fabricating new bodywork in 1988. The new Pirsch serial number for Truck 41 is #0044, with CFD Ship No. E-261. *Photo Credit: Bill Friedrich*

Serial No. IDA1P9CPD680MK004001: The last fire engine built in Kenosha under the Pirsch name was this Model CPD6 custom pumper destined for Osceola, Arkansas in 1991. Production of the truck was started during the Haas/Blondeck era some two years earlier. Kerry Poltrock, President of North Central Fire Apparatus, and his employees finished the truck. The pump is rated at 1,500 gpm with a 750-gallon booster tank. A 50-gallon foam cell was also included. *Photo Credit: Kerry Poltrock*

MORE TITLES FROM ICONOGRAFIX

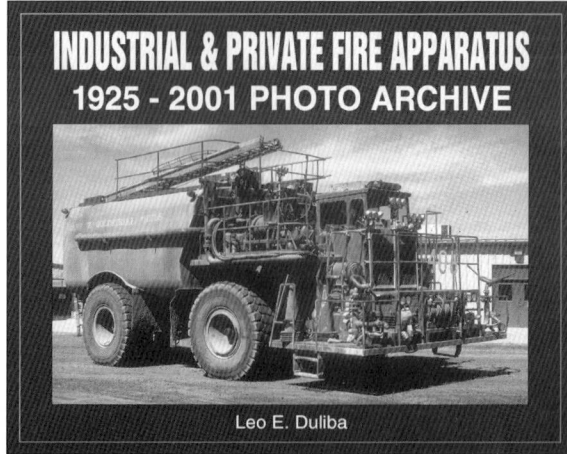
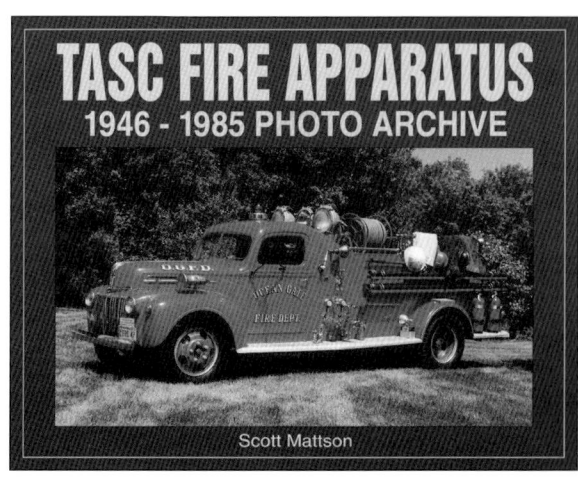